FEVR:
FORENSIC EVENTS VIEWED REMOTELY

Dr. Chuck Kennedy

FEVR: Forensic Events Viewed Remotely

© 2016 Dr. Chuck Kennedy

To the Remote Viewing Angels

At Project Psi Institute

You do almighty work

(blank)

Index

1. Introduction to Remote Viewing

2. A Short History of Remote Viewing

3. Remote Viewing Applications

4. Image Streaming: A Brain Exercise

5. Things That Will Enhance Your Session

6. How To Prepare

7. Setting Up The Case

8. The Process

9. Reaching Your Conclusion

10. Examining the Physical being

11. Examining the Mind

12. Glossary

(Blank)

1. Introduction to Remote Viewing

This workbook is to be used in conjunction with the FEVR Bootcamp or DVD Training; it is NOT a stand-alone textbook.

Remote viewing is a mental faculty that allows a perceiver (a "viewer") to describe or give details about a target that is inaccessible to normal senses due to distance, time, or shielding. For example, a viewer might be asked to describe a location on the other side of the world, which he or she has never visited; or a viewer might describe an event that happened long ago; or describe an object sealed in a container or locked in a room; or perhaps even describe a person or an activity; all without being told anything about the target -- not even its name or designation.

- When clairvoyance (RV's closest relative) was done under controlled conditions for research purposes, it was generally targeted at such things as cards or colors, since these sorts of targets allowed easy scoring of experimental results.

Remote viewing, on the other hand, was actually developed and first explored in a research setting (more about this in the history section). The sorts of targets used for RV research differed from those typically used in other psi research. Targets chosen for "viewing" include geographic locations, hidden objects, and even such things as archaeological sites and space objects about which it was expected that ground truth would eventually become known, so that the viewer's accuracy could be checked.

- Unlike most other psi disciplines, remote viewing is not precisely one thing, but rather an integrated "cocktail" of various phenomena. Despite the "viewing" part of the term, remote viewing is only partly about experiences associated with what might be visible about a target. It also involves mental impressions pertaining to the other senses, such as sounds, tastes, smells, and textures, as well as limited telepathy-like effects, and in some cases just plain intuitive "knowing." RV owes some of these qualities to the fact that lessons learned from research in clairvoyance, telepathy, and even out-of-body experiences -- traditionally considered separate disciplines -- played a role in its development.
- In remote viewing, the viewer not only verbalizes what he or she is perceiving, but usually also records in writing, in sketches, and sometimes even in three-dimensional modeling the results of the remote viewing episode, or "session."
- Remote viewing tends to be more structured than other psi disciplines. In some important varieties of remote viewing, viewers follow specific scripted formats. These formats are designed to enhance the viewer's performance in various ways,

such as to better deal with mental "noise" (stray thoughts, imaginings, analysis, etc. that degrades the "psychic signal") or to allow incoming data to be better managed. Some of these structural methodologies are widely used. Other methods are more personal. An individual remote viewer, for example, might through trial and error develop his or her own customized approach.

- Proper remote viewing is done within a strict science-based protocol. As mentioned, the remote viewer is kept unwitting of either the nature or identity of the target until after the session is completed. Except in training situations, the monitor (a sort of remote viewing "guide" or facilitator that may assist the viewer during the session) is also unwitting, and external clues or data about the target are carefully excluded. Sessions are conducted in a setting that prevents knowledge of the target "leaking" to the viewer. These measures are important to ensure that the viewer does not receive hints or clues about the target in any way other than what would be considered "INTUITIVE."

- Remote viewing is not used to give "psychic readings," "tell fortunes," "read auras," or other sorts of popular activities of this nature, but is rather a means of doing serious science research and for performing operational-type tasks in criminal investigations, government intelligence work, commercial applications, etc. Many who want to explore their individual human potentials also become interested in it.

Finally, one last point related to structures or methods that are often employed in remote viewing:

RV is not really a "psychic phenomenon" as such, but actually an imposed discipline or skill that helps the viewer to facilitate or "harness" his or her own innate, underlying intuition.

Some RV theorists think that formal RV methods are really just strategies that help the viewer to more successfully and reliably access the subconscious, where it is most likely that information obtained from RV first emerges into human consciousness.

2. A Short History of Remote Viewing

Remote viewing (RV) did not spring into existence overnight. Its earliest ancestors can be traced back thousands of years to the days of the early Greeks and beyond. But RV's most direct precursors date from the 1930's, beginning with experiments in clairvoyance under conscientious scientists like J.B. Rhine. Research into telepathy and "thought transference" by notables such as Upton Sinclair (described in his book *Mental Radio*) and Rene Warcollier (*Mind to Mind*), together with investigations into out-of-body states contributed further to developments that would eventually produce remote viewing.

In the late 1960's and early 1970's, out-of-body experiments were conducted in New York City by researchers at the American Society for Psychical Research. One of the subjects of these experiments was Ingo Swann, an artist and student of the paranormal who had come to New York years before from Colorado. Tiring of the standard research protocols, Swann suggested a number of changes and improvements to the experiments, which among other things led to a successful series of attempts to mentally describe the current weather in various cities around the US. After Ingo's descriptions, the weather conditions in these cities were verified by a phone call to a weather station or other reliable authority.

These experiments suggested to others that something unusual to current understanding was involved by the "remotely viewed" locations and objects otherwise inaccessible to direct human perception. The results were provocative and underscored the value of further research.

In 1972 Dr. Hal Puthoff, a physicist at SRI-International, a California-based research institute that had been spun off from Stanford University, expressed his interest to a researcher in New York in conducting research into a form of non-conventional communications. The New York researcher was an acquaintance of Swann's, which in fact eventually led to Swann and Puthoff getting together to conduct an experiment that ultimately attracted attention and funding from the Central Intelligence Agency. Research Physicist Russell Targ soon joined Swann and Puthoff at SRI, forming the core of a team that researched and refined understanding of what had now become known as "remote viewing." For the next two decades most remote viewing research was funded by the government and performed in secret. But a few less-secretive sources also provided support, and a limited amount of non-classified information about RV was published.

In the mid-'70's government support for the growing RV program moved from the CIA to the Defense Intelligence Agency, as well as certain other military organizations. Subsequent experiments and research explored the edges of what remote viewing could do and tried to improve quality and consistency of the results. In 1978, the US Army created a unit to use RV operationally in collecting intelligence against foreign adversaries. This program continued under Army sponsorship until 1986, when the operational and research arms of the government remote viewing program were combined under the leadership of DIA. In about 1991 DIA renamed the program "Star Gate."

By this time, the research part of the program had itself been transferred from SRI to the Science Applications International Corporation (SAIC), and was directed by Dr. Edwin May, who had replaced Hal Puthoff in 1985. Concurrent with the government RV program, civilian researchers were exploring phenomena related to remote viewing. Some of these were replications of SRI's experiments, while others followed complementary avenues of research. Most prominent of the latter were Charles Honorton's "Ganzfeld" techniques, and the "remote perception" experiments conducted at the Princeton Engineering Anomalies Research laboratory. Civilian applications were being explored as well.

(Blank)

3. Remote Viewing Applications

Why would someone want to remote view? What would they hope to gain? Well let's look at some of the more common applications. But in all honesty, the sky is the limit, literally. Common sense applications include:

- Corporate

- Education

- Futurtronics

- Law Enforcement

- Defense

- Healthcare

- Historical

Before entering into an arrangement in any of these fields make sure that it is clear what they expect of you. As you trust the universe to provide you answers to the mysteries of the ages, you must know that the universe has ethics and you will not be allowed to use your skills for anything unethical. So you won't be able to see next week's Powerball numbers. But you can find money that has been lost or stolen.

Many corporations will pay to do a deep background check on an applicant especially for an executive position. A resume and a reference may not tell the whole story. Many job references will only show dates of employment and if they are eligible to be re-hired. A viewer can see why they left the last job, as well as how well they got along with co-workers. Some companies have a big problem with employee theft. A viewer can not only spot who the thief is but also where embezzled money is. So those are just starters.

In Education, viewers can see what programs are on the drawing boards for education around the globe as well as what techniques stand the test of time, e.g. Common Core.

Futuretronics. What is that you say? Wouldn't you like to know what the next big tech trend is going to be? In the late 80's, Joe McMoneagle, one of the stars of Stargate, was asked by a tech company to predict the next big thing. He said that he saw kids walking around with tablets in their hand. The company officials laughed and shelved the idea. When the tablet market exploded that company was left in the dust.

Law Enforcement has many needs for remote viewers. In the past year, Project Psi has participated in 18 missing person cases.

When SRI started in the 1970s, it was estimated that to just take a wild guess at these things anybody would be successful 25% of the time. At Project Psi using a scoring system for accuracy, we have an overall success rate of 84.9%. For abductions, our success is 91.4%. I think that speaks for itself.

Defense is a huge need, and since 9/11 the need for security is very big. There always seems to be a threat of a terrorist attack.

Healthcare is always in demand of the newest drug, or medical miracle device. On an individual level the subconscious insertion is huge in assisting addiction issues.

Historically, wouldn't you like to be at Ford's Theatre on April 14th, 1865? Or July 21st, 1969 as the Eagle of Apollo 11 landed on the Sea of Tranquility? Or maybe just going to Arizona in 1892, to find the Lost Dutchman's Gold Mine. The infamous robber Sam Bass stole 3,000 gold bars. the gold is hidden, near Denton Tx. Every state has a history of hidden treasure So as you can see the uses are pretty much unlimited.

What would you charge for such services? Why don't you just RV to the people that want to hire you, and find out what they are willing to pay.

(Blank)

4. Image Streaming: A Brain Exercise

This entire instruction set is based on original works by Win Wenger, PhD as published in his ground-breaking book, *The Einstein Factor*, co-authored with Richard Poe. 1997

"Image Streaming" was developed by Win Wenger, PhD in the 1990's as a method of increasing brain activity in below par students. It turned out to be a spectacular success, but even more useful in already superior minds of freshman students enrolled in chemistry and physics courses in a small southwestern Minneapolis college. The instructor there performed the only scientifically-designed test of the efficacy of Image Streaming by measuring the before and after IQ scores of about 30 volunteers.

On average, the advanced-intelligence students' scores went up 15 to 20 points after 30 days of practicing the procedure.

This is what it is:

With your eyes closed, you audibly report what you see floating across the inside of your forehead.

What you perceive from your five senses is filtered through your limbic system (inner brain), sent through your prefrontal lobes of your brain (forehead) and thence into your cerebrum (grey wrinkled outer brain) for perpetual storage. You can see the information as it travels across your inner forehead as distinct images, blobs of color, or weird shapes.

When you command your vocal chords to describe what you see, the nerve system that translates your thought process into muscle movement (vocal chords) produces a sound. This acoustic energy is detected by both of your ears and translates into electrical pulses which are sent through to your left brain hemisphere which then sends it to your right hemisphere to be processed into a thought for filtering through your limbic inner brain to travel across your forehead up to your grey storage department. This process runs continuously while you are in the game.

This is a feedback loop. MRI testing shows that while in this loop, the brain lights up in all parts. This shows that brain activity is stimulated with no more than what I just described that covers all distant parts.

Each time a neuron is tickled to send the pulse (information) on to the next neuron, it grows a hard wire to match up with another neuron. What fires together, wires together. This goes on at the rate of about 200 mph. The wires grow very slowly. The signals travel at the high speed.

Twenty to thirty minutes of this activity is like body building with weights for a couple of hours. This is mental exercise unlike any other kind of mind work.

Images?

Why do you see images in the first place? The human brain stores images, not words or alphabet or numbers, but images. What you will see flashing across the tv screen inside your forehead are the images created in your limbic brain from the information flowing into it from the smell, sight, sound, touch, and taste sensors of your body.

The Process

Set a 20 minute timer on your phone so you aren't trying to keep track of time.

Quiet your brain to daydream level. You will be coasting at Alpha frequency.

Close your eyes.

Get comfortable. Take a deep breath.

Let it out slowly.

In through your nose, out through your mouth

Repeat this for 10 breaths.

Listen to your breathing.

Focus all your attention on just the sound of your breath.

Keep your eyes closed.

Then out loud describe what you see as you go around the room to each other item one at a time. Describe these items in as great detail as you can. Color, size, material, surface, and anything else (this will get the images flowing.)

Describe whatever you see, whether it's something you recognize, just a blob of color, or whatever.

The first time you try this you will feel like you have been talking for twenty minutes and it will only be three or four. Relax, and keep describing what you see. Your timer will stop you. Like anything you do in life, the more you do it the easier it becomes.

5. Things That Will Enhance Your Session

As you know many things enhance or restrict cognitive function. When viewing ideally you want to be in Theta, the brain state just before sleep. Some of the things that will help you get there include;

- Gingo Biloba

- Mugwort; either tea, tablet or sage

- Vitamin B6

- Chocolate; either sweet or dark

Coffee will enhance a viewing session, and alcohol will definitely impede viewing. Also, any central nervous system medication can affect how you perform; however, these medications affect everyone differently.

Antihistamine and cold or allergy medications are osmotic diuretics and can cause problems. Make sure you maintain proper hydration and sugar levels as these are critical. We will talk about these in depth in class.

So does who you are tell if you will make a good remote viewer? It may just contribute.

One study found that the best remote viewers are:

- Postmenopausal Women between 50 and 65

- Gay Men

- All other Women

- All other Men

6. How to Prepare

Set aside about an hour of uninterrupted time.

Start your PC; Pre-Conditioning.

That is 15 minutes of meditation to clear your head.

Put on your headphones with your binaural beats; Theta-state music

That comes with the program, to get you ready.

When you feel relaxed and in Theta, start your header.

(Blank)

7. Setting Up The Case

Get your Case ID number from your tasker or create one if you are working alone.

Start with a plain sheet of paper. And start.

The Header

The header section is the administrative page of the protocol.

Step 1: Place your name in the upper right-hand corner of your paper.

You can choose to use your given name, or your code name.

Step 2: Immediately below your name, write the date, including the date, month, and year.

Step 3: Under the date, write the time you are beginning the session. Please add your time zone.

Step 4: Place the Case Identifier in the upper left-hand corner of your page, prefaced by the abbreviation CASE ID:

Step 5: ES Emotional state. Drop pen.

Step 6: PS Physical state. Drop pen.

Step 7: Draw a line.

Step 8: Write Case ID

Step 9: Let your hand go.
Doodle whatever comes to your mind for 10, 20, 30 seconds then stop. Grab a fresh blank page to start your first scan.

8. The Process

The Scan

After you number the page (page 2) and write the "S" at the top center of your page to signify that you are doing a Scan, your next job is to write the Case ID in the upper left-hand corner of the page.

Immediately following the ID, at the very moment you end the last number or letter, let your hand do a quick, small, automatic doodle. This doodle is the secret to the universe. It holds all the information you need about your objective.
This should take you no more than 3 seconds.
There should be no hesitation between the last of the tag and
the beginning of the doodle.

There are not many directions you could allow the pen to travel while it makes this doodle. The shape that the doodle takes tells you some basis gestalt level descriptions:

- If the line is curved, like a hump, then most likely your objective has something mountain-like in it.
- If the line is wavy, like the ocean, then most likely your objective has something fluid like water in it.
- If the line is straight and unwavering, then most likely you

are remote viewing something with a flat surface.
- If the line is circular, making a loop, then most likely you are viewing a being that holds some kind of animation or sentience.
- If the line is chaotic, vertical, and wild, most likely you are viewing something that has energy or kinetics.

I use the term "most likely" in describing these doodles, because every objective is unique, and every viewer is unique, and we are by nature exceptions to the rule. But generally, those are the basic gestalts that you, as a new viewer, will encounter.
The term used in Remote Viewing for these doodles is "**ideograms**."

Basic Ideogram Shapes:

subject
(i.e. sentient being)

water (or liquid)

energy

flat surface

The ideogram is an amazing beast. It holds the taste, color, temperature, emotion, and activity of your objective in such a small package.

Now that you've drawn an automatic doodle, an ideogram, it's time to let the ideogram tell you something about the nature of your objective.

You move your hand to the other side of the page, to the right-hand side. Write the letter "A" followed by a colon.

Next, take your non-dominant hand and trace over your ideogram, as if you were recreating the motion your pen made with your finger. As you are tracing over your ideogram, use your dominant hand and write down the motion that your finger is making using words. You will want to use words that show direction and action, such as "sloping," "looping," "angle upwards," etc.

Here is an example of how this may look:

1/20
1601

A: Wavey across.

When you put the pen to the page, you probably don't pay attention to how the paper feels beneath your touch. If you think about this, you probably expect the pen to press into the paper, into the desk underneath the paper, the result being a bit of hardness in the process. But that isn't what happens.

Something about putting your attention to discovering certain kinds of data, answering certain questions, causes you to walk through a portal where the laws of physics are suspended.
After you've written the description your pen has made while creating your ideogram, the next step is to feel your ideogram to find out the physical density of this aspect of your objective.
You do this by taking your pen and literally pressing into your ideogram. You can press into it once, or twice, three times, whatever you need. As you press your pen into the ideogram, ask yourself **"How hard does this feel?"** and you will get an answer by how the pen feels as you push it into the page.

It might feel like it is meeting a lot of resistance, that the surface is hard. It might feel soft, as if your pen is being pushed into a stack of down pillows. It might feel mushy, as if your pen is pressing into wetlands. Whatever the physical density is, even if it seems ridiculous, write it down underneath the description of the line motion. This is still under part A:, still part of decoding your ideogram.

Viewers are so different. Some are extremely kinesthetic, and these viewers will have no problem literally feeling the resistance against the pen. Others are more conceptual, and won't feel a literal density, but once they press their pen into the ideogram shape, they will just Know what the density is. It doesn't matter which kind of viewer you are, you will be completely correct in your decision.

Your page will now look something like this:

You can see that this viewer determines the density to be "firm, mushy."

At this point all of your data needs to be "basic" concepts. If you are getting a vision of a person or place this early in the process

you need to declare it as an AOL; an Analytical Overlay. When this occurs, just write AOL and describe what you are seeing.

Next, you take your non-dominant hand, and feel your ideogram. This time you ask yourself **"What is the shape or surface curvature of this aspect of the objective?"**
This is called sensing your ideogram for the topology. Topology is just a fancy word that means the shape or contour of any object or surface. Just like pressing your pen into the ideogram to determine the physical density, when you use your hand to feel for the shape or contour, you may literally feel your hand make the motions of running over the actual shapes that are at your objective. Or, you may just Know what the shape or contour is. Either way, write your perception of the topology down underneath the physical density.

Your page will now look something like this:

$$S \qquad 2$$

1120
1601 ～～

A: Wavey across.

Firm
But Mushy
level

You can see that this viewer determines the topology to be "Level."

Part B:

The next part of your scan is where you combine your intuition and your common sense and make a decision about this aspect of your objective.

Write the letter "B" with a colon after it underneath the place where you wrote the topology.

Then take a good look at your ideogram. Remember what it felt like, the curve and heaviness of it. Declare what your ideogram represents. Use low-level language, such as "subject, mountain, structure, water, flat surface, energetics."

After you declare what it is, you then think about it for a moment and describe another attribute of the objective, something along the lines of whether it is manmade, artificial, natural, moving. Keep this description simple.

You don't want to decide exactly what your objective is, yet. You want to be describing it, using simple terms. You can describe anything in the universe using simple terms and if you use enough simple terms of varying kinds, the person reading your paper will be able to visualize what you are describing.

Part B will look like this:

```
                 S                          2
  1120
  1601    ~~~
                      A: Wavey
                         across.
                         Firm
                         But Mushy
                         level

                      B: Water
                         moving
```

The viewer declares that this ideogram represents water.

Part C:

Next, move to the left side of your page. Underneath your ideogram, write a "C" followed by a colon.

Write a quick stream of sensory impressions, things like:

- Colors (What colors does probing your ideogram generate? Red? Blue? Aquamarine?)
- Sounds (What sounds do you hear? Birds chirping? Sizzling? Tick-tock?)
- Textures (What textures do you feel? Gritty? Smooth? Soft? Fuzzy?)
- Temperatures (Is it hot? Cold? Warm?)
- Movements or Energetics (Is there anything moving? Is it static?)
- Shapes and Surface Curvatures (Are there circles? Squares? Ridges? Edges?)
- Smells (What smells do you perceive from your ideogram? Acrid smoke? Sweet perfume?)
- Tastes (What do you taste? Oil? Fish? Cigarettes?)
- Relative Dimensions, such as tall, short, thin, fat, etc.
- ANYTHING having to do with the five senses.

You should have at least five impressions here, but as in most things in life, more is better!

You should probe your ideogram, press into it with your pen, in order to gather the data. Or you can take your non-dominant hand and run it over your ideogram.

The more you involve your body, the better. You can place your hands in front of you as if you were massaging a model of the objective and run your hands along invisible sides and into unseen corners. You can stand up to get a sense of size. You can put your pen down for a moment to gather your thoughts.

Keep in mind that you should be using simple terms here. You don't want to name things yet. You want to use rich description so that the person reading your session later will be able to build a mental impression of the objective.

Part C will look like this:

S   2

1120
1601 ~~~

A: Wavey
   accross.

   Firm
   But Mushy
   level

C: Cold, wet
   Noisey
   Salty
   fishy
   massive

B: Water
   Moving

Part D:

After you feel that you have exhausted the intuitive stream, write a "D" followed by a colon in the remaining space on the page. Underneath the D: you will sketch this aspect of the objective. Before you begin, review your data. If you have described a tall, rounded, elongated structure with something moving below it, then this is what you will draw.

You can close your eyes and draw what you see in your mind's eye. Some viewers are more visually-oriented than others, so if you are not one of these individuals, then you can set the tip of your pen to paper and allow your hand to move over the page by itself, without conscious urging.

Regardless of the method you choose to gather the visual data, it is imperative that what you declare in part B is what you draw in part D: You should label sections of your sketch to indicate motion, textures, color-- any data that is relevant.

5 2

1120
1601 ~~~

A. Wavey
across.

Firm
But Mushy
level

C: Cold, wet
Noisey
Salty
fishy
massive

B: Water
Moving

D: ~~~~
    ~~~~
    ~~~~

Three Times a Charm

You should do three scans of your objective. Each one will be unique, and will describe a different aspect of your objective. You may get different ideograms each time, or you may have an ideogram that repeats. Follow your body and your intuition in each scan, and don't assume that scans must match each other.
For example, if the case was the sinking of the Titanic, one scan may be water, the second could be a structure--the ship. Then, the third could be a sentient being, either a survivor or a casualty. In this case, each scan would be completely different from the others.

The Collector

After you have completed the three scans which comprise the Survey section of the session, you move on to the Collector. You start this section with a clean piece of paper, in portrait mode - taller than it is wide. Put your page number in the upper right-hand corner. You should be at page 5 if you have done three Scan pages plus the Header page.

There are two types of Collectors in our system of Remote Viewing. You only use one type per session. The most common of the two types is the Open Collector. The other type is the Defined Collector. Each is used for different reasons, as outlined below.

The Open Collector

The Open Collector is used when you want to integrate all the data together achieved thus far in a session.

The scans allow you to look at individual aspects of your objective, so by the time you have completed three scans, your subconscious mind has isolated bits and pieces of the objective, but has not put them together into some kind of a whole. The Open Collector begins putting these elements together so that your mind may start making sense of the objective

The Open Collector also allows the viewer to begin experiencing emotive, psychological as well as low-level conceptual data.

In the upper-middle portion of your page 5, write the abbreviation "OC". This stands for Open Collector.

Take your three scan pages and fan them out to the side of your non-dominant hand, so that the part D sketches are showing. Run your non-dominant hand over the sketches. Write down a stream of sensory data, adding emotive, psychological, and low-level conceptual data as it occurs.

In a way, this is much like the part C in the scans, but you are now gathering data that describes the objective in full, not-isolated aspects.

It's alright to repeat some of the data you have already used in part C of previous pages, but you should not be copying these descriptions. Let the perceptions rise out of the page as you run your hand over the sketches. You may string descriptions together into phrases if it occurs naturally.

Strive to fill your page with perceptions. It helps to keep running your non-dominant over the fanned-out sketches.

Here is an example of an open collector

1 Feb 2012 — OC

- broken up
- scattered
- in pieces
- Nostradamus
- swooping through
- rounded
- pancake
- my hero
- unusual
- Nostradamus
- swaying
- immense
- unfounded
- unusual
- carrot cake
  → is a lot of pot-luck items seem to be appearing!! all round! (pizza pie, pumpkin pie)
- a lot of people gathering
- fighting
- singing
- large gathering
- uncounted / uncountable
- uncontrolled
- enhancing
- St. Peter
- advancing

- adherent
- observer
- mistaken
- unusual
- fanciful
- drastic
- Satan
- abnormal
- unhearing
- unseeing
- unfounded
- environment
- unusual
- unblasted
- un, un, un'
- unaccommodating
- swirling
- dancing
- singing
- merriment
- obnoxious
- deterrent
- abstract
- defining
- enhancing
- mistaken
- unusual
- environment

- be rounded
- behemoth
- monster
- dandi
- dandelion
- mechanic
- monster
- betrayal
- denial
- hatred
- misery
- despair
- The Truth
- the hound dog
- hounded
- mistaken
- flogged
- denial
- mistrust
- mistreatment
- abuse
- uncounted
- unassuming
- mistreatment
- manhandled
- despair

If you are unsure of which type of Collector to use in a session, and the tasker has not instructed you to use either Open or Defined, most viewers choose to use an Open Collector.

Defined Collector

The Defined Collector is used when you want to keep the aspects described in the scan pages separate. This can be extremely useful at times when your scans seem disjointed in some fundamental way. Sometimes the person assigning the session to you will tell you to do a Defined Collector.

When you use the Defined Collector, you are instructing your subconscious mind to continue to keep the central aspects described completely separate.

In the upper-middle portion of your page 5, write the abbreviation "DC". This stands for Defined Collector.

Take your first scan and run your non-dominant hand over the part D sketch. On your DC page, write a number "1" on the left-hand side and write a column of sensory, emotive, psychological, and low-level conceptual data that relates to the sketch in scan 1.

When you have completed your column of data, take your scan 2 page, and write a number "2" in the center of the top of your Defined Collector page and then, running your non-dominant hand over the sketch in part D of scan 2, write a column of perceptions relating to this aspect of your objective.

Repeat this once more for scan 3. Write a number "3" in the right-hand side of the page, and fill a column of data relating to the aspect of the objective in this scan.

You should have roughly twenty perceptions written for each column in a Defined Collector. More is always better!

Here is an example of a Defined Collector:

DC

5(1)

- cobblestones
- wooden wheels
- brown
- pinkish red
- sack cloth
- rough texture
- walking
- bare feet
- smiling
- flushed cheeks
- chapped faces
- cold
- green
- rest stop
- feminine
- moving from then to now
- well traveled road
- truth search
- gender unimportant
- ladies apron
- long dresses

5(2)

- dead end
- {D: uncertainty past will}
- feminine figurehead
- white
- flowing garment
- brown skin
- necklace or rosary
- wood cross
- beautiful
- poor
- spiritual quest (knowledge)
- natural living
- peaceful scenery
- daylight
- spinning top
- string - piece of wood at end
- esoterical
- abrupt stop (end)

5(3)

- glass
- shiny
- rose red
- edges smooth
- wings
- insignia
- transparency
- swing
- straw
- organza fabric
- cool structure
- smooth feel
- smallish
- meaningful
- icons
- representative
- facade

5(1)
S: moving figures unimportant; action matters; teaching, learning, leading, following

The Grand Sketch

After you have completed your Collector, you move on to the Grand Sketch.

You start this section with a clean piece of paper, in landscape mode - wider than it is tall. Put your page number in the upper right-hand corner. You should be at page 6 if you have done three Scan pages plus the Header page plus the Collector.

In the upper-center of the page, write the initials "GS", which stand for

"Grand Sketch."

The Grand Sketch is where you combine all the aspects of your objective into a comprehensive, detailed sketch. You should allow yourself some time to put these pieces together. The Grand Sketch should never be rushed.

So far in the session, the sketches drawn have been simple and intuitive. The Grand Sketch needs to show detail and purpose.

Review all the data you have written up to this point. First consider your scans. Take the elements you have already drawn in your scans and find where they belong, and in what relationship to each other, in your Grand Sketch.

Begin sketching, combining aspects together, adding new elements as they arise, making sure to label any movement or activity.

You can close your eyes during this process in order to see images. You can keep your eyes open and draw what you intuitively and intellectually know is present. Follow the sketching method that makes the most sense for you.

Here is an example of a Grand Sketch:

9. Reaching Your Conclusion

The next section of the session is called the **Matrix**.
Until this point, the viewer has been getting simple impressions of what is occurring at the objective. There hasn't been a way to consolidate and melt data together in a way that enlightens the viewer. The Matrix section is designed to teach the viewer to look for different types of higher-level data while forming more fully-conscious thoughts about what is present at the objective.
The Matrix looks much more complicated than it is. You begin with a piece of blank paper in landscape mode. Put your page number (typically page 7) in the upper right-hand corner of the page. Place a capital "M" in the center-top of the page.
Along the top of the page underneath the "M", you write this string of letters:

S M T E Sj P Pp C SB PR/V

Each of these letters stands for a different type of data. Underneath the string of letters, you draw a line across your page.
The top of your Matrix page will look like this:

S   M   T   E   S_T   P   P_p   C   S_b   PR / V   P_7
                                                    M

Let's go through the definitions first.

"S" stands for "sensory" data.
Sensory data is all the data that pertains to the five senses, such as colors, textures, smells, tastes, sounds, and temperature. Some examples of sensory data are:
Hot, Red, Gritty, Chirping, Oily-smell, Bitter

"M" stands for "magnitudes."
Magnitudes are 'how much' or 'how many' of something, including relative dimensions. Some examples of magnitudes are:
Tall, Fat, One, 12 Dozen, Few, Some

"T" stands for "topological" data.
Topology means the surface curvature of an object or place. If you run your hand along the surface of something to feel the shape or the curve of it, that would be the topology. Some examples of topology are:
Square, Oval, Ridged, Pointy, Triangle, Beveled

"E" stands for "energy" data.

Energy data are the data that pertain to any type of action or change, including types of energies. Examples of energy data are:
Vortex, Spinning, Electromagnetic, Biochemical, Sliding, Hopping

"Sj" stands for "subjectives."

Subjectives are the emotions or psychological states of the people, places, and things at the objective. These are NOT the viewer's emotions. Some examples of subjectives are:
Love, Peaceful, Anger, Concern, Sadness, Schizoid

"P" stands for "physicals."

Physicals are all people, beings, places, and things in 3-D. Anything you can touch in some way is a physical. Some examples of physicals are:
People, Wood, Structures, Water, Mountains, Objects

"Pp" are "paraphysicals."

Paraphysicals are all people, beings, places, and things in non 3-D. Anything that is a real thing but you cannot touch in any way, is a paraphysical. Some examples of paraphysicals are:
Some aliens, Ghosts, Angels, Massless particles such as tachyons, Waveforms

"C" are "concepts."

Concepts are all idea words-- words that stand for something but have no physical presence. Some examples of concepts are: Knowledge, God, Library, Desire, Format, Magic, Expression

"Sb" are "symbolics."

Symbolics are symbols that stand for something else. Sometimes these are visual symbols such as a Star of David or a peace sign. Sometimes these are proper names such as "Bob" or "Katy," or a number out of the blue.

"PR" are "personal reactions."

Personal reactions are the feelings and emotions that the viewer is having. These are NOT the feelings and emotions present at the site of the objective, though they may be similar to what people at the objective feel. For example, if you are viewing a disaster such as the sinking of the Titanic and you are feeling upset by what you are viewing; being upset would be your personal reaction.

"V" are "visuals."

Visuals are small snippets of visual data that directly relates to what is present at the objective. These are small sketches.

MATRIX RUN

Start under "S" and probe for some sort of sensory data, then drop down a line and move over to "M" and probe for magnitudes. The reason we drop down is to show the connection of the data, and it shows a timeline.

You will want to do a minimum of two Matrix runs.

Now pull out your sketches and choose the one that appeals to you the most. You have now put on your detective hat and gotten out of the intuitive mode. Pick an element of the picture and probe for one of the six gestalts-- ask yourself detective questions; is it hard or soft,

big or small, the shape, etc. At this point you might feel you can describe what it is so you can declare an **Intuitive Statement.**

Write "IS" and describe what you see, then probe to see if it is static or moving.

Probe all around and find out everything you can about the sketch. It is difficult to probe for gender-- you literally reach out and grab for gender parts. Observe young or old age by what they wear and what they do for a living. If you feel one area is key on your matrix, focus on that key and your subconscious will shift to the new perspective. Usually at this point the data comes out like IS data in full streams.

At this point look around the scene in a 360° panoramic, walk around the scene to get a better angle, or take yourself 1,000 ft above the scene and using the scene as the center of a clock, describe what is at 12:00, 1:00, 2:00, and so on. Identify the distance from the scene.

You can morph with the structure or blend with the subject if it was a traumatic event. DO NOT BLEND WITH THE PERSON. IF YOU DO BLEND and you are uncomfortable just write "End of Blend" and drop your pen. As you run the Matrix, your case will come into Focus.

10. Examining the Physical being

How to Discover about your Subject's Physical, Spiritual, Psychological and Emotional Bodies.

Physical Profiling is done anytime you have a subject in your GS. Go to your "Physical Profile" page. Probe the body; hair to get the hair color, probe the eyes and get eye color, etc., and write it right on the PP sheet, or the Matrix. Probe everywhere to get the condition of each organ. The outer ring is the sphere of influence, or the aura. A lot of somatic issues are a result of emotional issues. The more you probe the more you get. You probe the brain or heart and then ID alive or dead, or if they have artificial limbs. You can also pick up jewelry or tattoos.

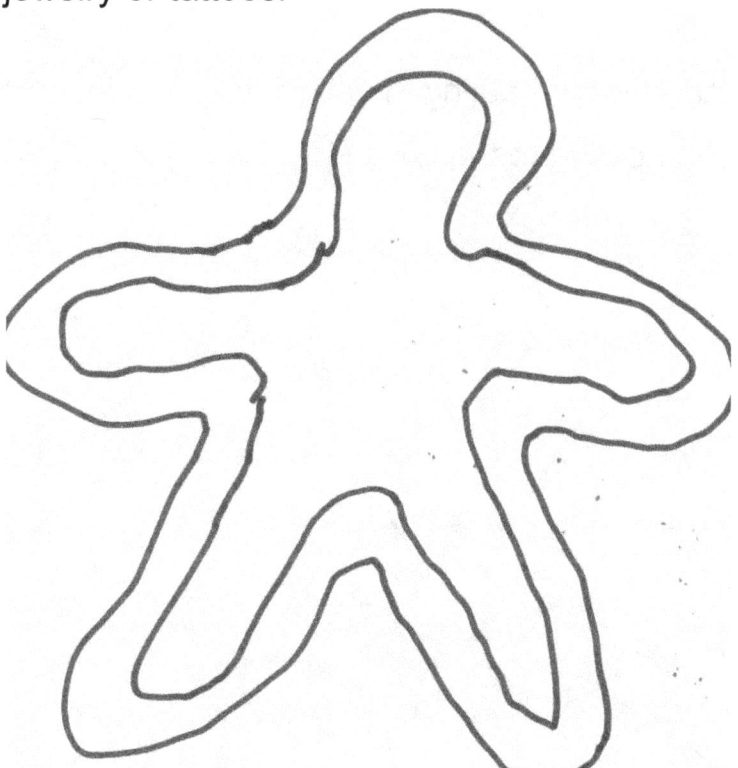

11. Examining the Mind

How to do a Consciousness Map:

Draw a circle with an "X" through it, then draw four columns; "CC", "CT", "SC", and "ST."

CC; Conscious Condition-- how they feel on the conscious level.

CT; Conscious Thoughts.

SC; Subconscious Condition--e.g., lonely, self-esteem, worried, insecure, religion styles, etc.

ST; The thoughts you tell yourself-- e.g., "I have to get out of here", "I'm tired", "I gotta pee", etc.

You put your pen in the center of the X and pull out the thoughts; they are usually in the first person. The conditions are usually feelings.

To heal, you write the healing intended, then circle the healing and direct the healing to the ST line.

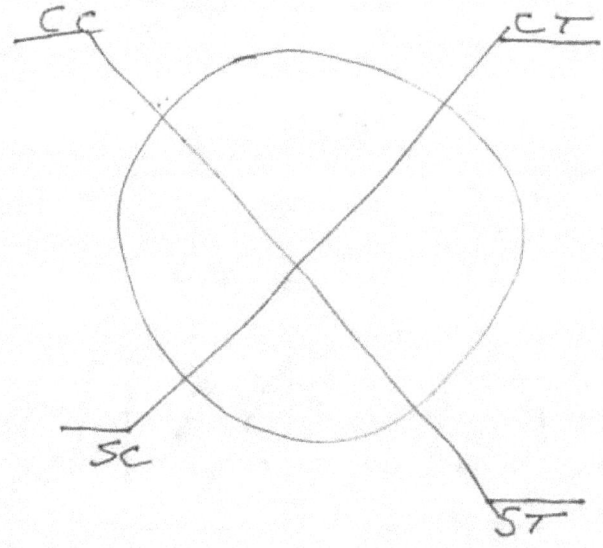

12. Glossary

Analytical Overlay (AOL): A graphic, big-picture vision of a case early in the process.

Concepts: All ideas and words that stand for something but have no physical presence.

Consciousness Map: A map of the conscious and subconscious of your subject.

Defined Collector: Used when you want to keep the aspects described in the scan pages separate.

Energy Data: The data that pertain to any type of action or change, including types of energies.

ES: Emotional state.

Grand Sketch: Where you combine all the aspects of your objective into a comprehensive, detailed sketch.

Ideogram: A written character symbolizing the idea of a thing without indicating the sounds used to say it.

Magnitudes: Pertains to amount; 'how much' or 'how many' of something, including relative dimensions.

Matrix: The section is designed to teach the viewer to look for different types of higher-level data while forming more fully-conscious thoughts about what is present at the objective.

Open Collector: Used when you want to integrate all the data together achieved thus far in a session.

Paraphysicals: All people, beings, places, and things in non 3-D. Anything that is a real thing but you cannot touch in any way, is a paraphysical.

Personal Reactions: The feelings and emotions that the viewer is having.

Physical Profile: A body scan of anyone key to the case when they come up.

Physicals: All people, beings, places, and things in 3-D. Anything you can touch in some way is a physical.

Probing: Poking at your ideogram with your pen to get a feel for what is there.

PS: Physical state.

Psisomatic: A tool for allowing yourself to be free of your

conscious mind.

Sensory Data: All the data that pertains to the five senses, such as colors, textures, smells, tastes, sounds, and temperature.

Subjectives: The emotions or psychological states of the people, places, and things at the objective.

Symbolics: Signs and images that stand for something else. Sometimes these are visual images such as a Star of David or a peace sign.

Topology: The surface curvature of an object or place. If you run your hand along the surface of something to feel the shape or the curve of it, that would be the topology.

Visuals: Small snippets of visual data that directly relates to what is present at the objective.

NOTES

www.ingramcontent.com/pod-product-compliance
Lightning Source LLC
Chambersburg PA
CBHW081255180526
45170CB00007B/2430